UFOs

"The Official Government Acknowledgement of Meetings with Alien Visitors"

ARTHUR ELLINGTON

Copyright ©2023ArthurEllington

All rights are reserved. Except for brief quotations used in critical reviews and certain other noncommercial uses allowed by copyright law, no part of this publication may be duplicated, distributed, or transmitted in any way, including by photocopying, recording, or other electronic or mechanical methods.

The Official Government Acknowledgement of Meetings with Alien Visitors

Table of Contents

INTRODUCTION

Remarkably, 95% of Americans have heard of or seen references to unidentified flying objects (UFOs), and a startling 57% of them think they are real. Even the late US presidents Carter and Reagan claim to have seen a UFO. The nation is home to a network of fervent UFO enthusiasts who go by the name of "UFOlogists," who pursue UFO-related interests. Notably, a great number of people sincerely believe in a massive government cover-up pertaining to UFO investigations, especially by the CIA. Since the late 1940s, when the contemporary UFO phenomenon first emerged, UFO enthusiasts have continued to hold a primary narrative that involves the CIA's covert concealing of UFOs.

From ancient cave drawings of strange animals falling from the sky to modern eyewitness accounts of flying saucers and mysterious lights, the mystery surrounding UFOs has persisted throughout nations and civilizations.

For a long time, sightings of UFOs were dismissed as myths and folklore, either as pure imaginations or as the result of people's overactive imaginations. Governments and the scientific community often discounted reports of flying saucers and other unexplained aerial occurrences, dismissing them as flaws in natural phenomena or covert military operations. Because of this, there was a sense of mystery and suspicion surrounding the subject, which fueled conspiracy theories and gossip.

However, the curtain of secrecy has begun to open recently, and official government positions around the world have changed dramatically. This is the backdrop for the gripping narrative "UFOs: The Government's Official Acknowledgment of Alien Visitors."

The Official Government Acknowledgement of Meetings with Alien Visitors

This book takes us on a journey through UFO sightings and the government's response to these unexplained encounters. Reviewing old chronicles and manuscripts that contain reports of close experiences with the weird will be part of our journey through the history of UFOs.

We'll examine the intriguing tales, ranging from the "flying shields" era in ancient Rome to the mysterious airship sightings of the late 19th century, that raise the possibility of interactions with advanced civilizations outside of our solar system.

As we proceed, the tour's itinerary will drastically alter to take us to pivotal moments in contemporary history when official government investigations into UFO encounters were launched. One such historic project was the American Air Force's Project Blue Book, which was tasked with looking into hundreds of UFO cases in the 1950s and 1960s. We'll look at the highs and lows of this significant initiative and the conversations it spurred.

Our fascination and curiosity increase as we move into the shadows, where top-secret government research on UFOs is kept out of the public eye. Although the results of these clandestine projects have been kept under wraps for many years, we will attempt to make some of them public and shed light on the covert studies into the UFO phenomenon.

The tide turned because reliable witnesses and persistent researchers refused to be intimidated by the government's denial and mockery. The emergence of whistleblowers and leaked documents created new avenues for inquiry, leading governments worldwide to demand increased transparency.

Our journey comes to an end with the discovery of an unprecedented shift—an acknowledgment that would alter the course of human understanding.

Officials from countries that had previously maintained a low profile regarding unidentified flying objects have abruptly revealed information from their archives and admitted to having had close encounters with intelligently controlled objects of unknown origin. We'll look at how this revelation changed people's perceptions of the universe and how it affected society as a whole.

As we navigate this gripping tale of unexplained flying objects, let's embark on an investigation and discovery journey to uncover the truths that lie beneath the surface. We hope to shed light at the end of the journey on the significant implications of the government's formal recognition of alien visitors, as humanity stands on the brink of a new era in which the mysteries of the cosmos may soon be revealed to us.

CHAPTER ONE

AN ANCIENT MYSTERY RELATING TO UNIDENTIFIED FLYING OBJECTS

As we explore the murky corridors of history, we come across interesting clues that suggest UFOs may have existed in ancient civilizations. Our ancestors left behind a trail of enigmatic artefacts that have piqued interest for millennia, such as the vibrant cave paintings of figures that resemble extraterrestrials in Val Camonica, Italy, which date back over 10,000 years, and the intriguing winged discs depicted in hieroglyphs on the walls of Egyptian temples. These enigmatic reports of "fiery discs" and "celestial chariots" suggest that humanity's fascination in alien life is a long-standing tradition.

The details of what private pilot Kenneth A. Arnold witnessed on June 24, 1947, when flying above Mt. Rainier seventy-five years ago remain unknown to us. Millions of people worldwide now use the term "flying saucer" in their vocabulary because of what he claimed to have seen and spent the rest of his life attempting to explain.

Arnold left Chehalis, Washington that June afternoon for an aviation show in Pendleton, Oregon, stopping in Yakima, Washington, for fuel. In addition to being a member of an Idaho search and rescue team, he had recorded 4,000 hours of flight time as an experienced pilot. He was flying a small aircraft, a CallAir A-2, with a single engine. The winds were mild and the skies were clear. On the way, he intended to take a little diversion. Somewhere along his eastward route, a U.S. Marine Corps Curtiss C-46 Commando transport had crashed with 32 U.S. Marines aboard; Arnold planned to locate the downed aircraft and collect a $5,000 prize.

Arnold observed a brilliant light to the northeast just before 3:00 p.m. while circling his aircraft approximately 20 miles west of Mount Rainier in pursuit of the C-46. It surprised me. I just thought that it was a military lieutenant who was out flying a sparkling P-51, and that I had captured a reflection of the sun on his plane's wings. Arnold ruled out a nearby Douglas DC-4 aeroplane as the cause as further flashes emerged. He said that they came from nine bright objects in an echelon pattern that was around five kilometres long. According to Arnold, each object was round, roughly 100 feet wide, and lacked a tail. Periodically, the items wove, banked, and flipped side to side—"like the tail of a Chinese kite."

Arnold chose to schedule the formation's movement from Mount Rainier to Mount Adams since it was passing in front of him. He estimated that the objects were travelling at a speed of around 1,200

The Official Government Acknowledgement of Meetings with Alien Visitors

mph (other reports claim 1,700 mph), which was double the speed of any known aircraft at the time. It would take months before Col. Chuck Yeager broke the sound barrier and reached 700 miles per hour in the Bell X-1 rocket aeroplane.

Though Megan Garber noted in her June 15, 2014, piece for The Atlantic, "Stories of the time credit Arnold with using the terms "saucer," "disc," and "pie-pan" in his description of the objects he had seen," Arnold vehemently disputed having originally referred to the objects as "flying saucers." The day after his encounter, he reported his account to reporters Bill Bequette and Nolan Skiff of the East Oregonian newspaper. Skiff released a brief print piece the same day and used the phrase "saucer-like aircraft." Arnold was advised by Bequette that a wire story could prompt the military to comment on experimental aircraft flights, which could provide an explanation for Arnold's sighting. The Associated Press wire service published a brief story that Bequette suggested Arnold read, referring to what Arnold claimed to have seen as "nine bright saucer-like objects." The story that he had witnessed "flying saucers" had gone viral by afternoon throughout the country. When speaking with Arnold on a radio programme on June 26, the presenter observed how quickly the news spread, stating, "The Associated Press and United Press, all over the nation, have been after this story." Every newscast, radio programme, and newspaper that I'm aware of had mentioned it. The Chicago Sunran also the title, "Supersonic Flying Saucers Sighted by Idaho Pilot." Arnold rose to fame in the media, but he was not amused by the attention. "I have, of course, suffered some embarrassment here and there by misquotes and misinformation" that have been published in different places, Arnold said in an interview conducted thirty years later.

In many ancient tales, gods and goddesses descended from the sky, wielded advanced technology, and imparted wisdom to human civilizations. Hindu texts describe

The Official Government Acknowledgement of Meetings with Alien Visitors

"Vimanas"—flying devices that can traverse enormous distances in the heavens—while the Sumerian "Epic of Gilgamesh" tale describes the hero Gilgamesh's heavenly beginnings. These ageless tales, ingrained in the culture of bygone eras, pose a challenge to our understanding of the boundary between myth and fact.

Strange Aerial Encounters in Historical Accounts

Over time, accounts of unexplained aerial occurrences from a variety of civilizations changed. During the Middle Ages, sightings of strange flying objects were interpreted within a religious context. A classic example is the 1917 "miracle" of Fatima, when three shepherd children in Portugal claimed to have seen a bright apparition dance in the sky. This occurrence was seen by a large number of

The Official Government Acknowledgement of Meetings with Alien Visitors

individuals, which led to a heated debate on the nature of the encounter. While some believed it to be a supernatural visitation, others speculated that it may have come from another planet.

Throughout the late 19th and early 20th centuries, local inhabitants were terrified by tales of mysterious "airships" soaring through the sky. Witnesses claimed to have seen people that looked human or extraterrestrial piloting dirigible-like craft or moving forward with mysterious abilities. The Great Airship Wave of 1896–1897 in the United States sparked worries about advanced technology and the possibility of alien visits, making it one of the most perplexing periods in UFO history.

Sightings of UFOs in Europe

- 1942–1945: During bombing attacks over Germany, Allied pilots reported seeing strange metallic discs and balls of light. The

Allies thought the so-called "Foo Fighters" were a new Nazi weapon. It was discovered after the war that German aviation crew had also observed them but had no idea where they had come from.

- 1946: Over Finland, Sweden, and Norway, strange objects were seen. Reports of seeing these "ghost rockets" attracted the attention of both US and UK intelligence, but an explanation was never discovered.

- 1956: In August, there were many reports of UFO sightings over England. Two Venom fighters were scrambled by the Royal Air Force (RAF) to attempt to intercept an item that was being detected by radars at RAF Bentwaters and Lakenheath. A pilot reported seeing the UFO evading the aircraft.

- 1980: In December, members of the US Air Force stationed in England at RAF Bentwaters and RAF Woodbridge reported witnessing an unidentified flying object land in Rendlesham Forest, Suffolk. The landing location was found to have radiation levels 10 times higher than the surrounding average.

- 1981: In January, a UFO was seen touching down close to the French commune of Trans-en-Provence. At the landing location, strange markings were discovered, and a soil study revealed many irregularities.

- 1990: On November 5, a UFO passed several swift jets that were entering Dutch airspace on their way from the UK to Germany. The British Ministry of Defence received an official report from the pilots.

- 1993: Military troops saw a UFO fly over RAF Shawbury and RAF Cosford in March. The UK Ministry of Defence received a report.

The Second World War and the Development of Contemporary UFO Sightings

Many significant technical advancements occurred in the 20th century, such as the invention of jet propulsion and radar systems during World War II. As the number of artificial aircraft in the sky rose, so did the number of reports of unexplained objects. Pilots were often confused by fast-moving, unexpected airborne phenomena that seemed to defy established flying regulations, particularly in times of conflict.

Two years remained in the World War II. However, it seemed more like the start of the War of the Worlds to the pilots of the 415th Night Fighter Squadron.

The first person to view it was Lt. Fred Ringwald. In a night fighter flown by Lt. Ed Schlueter, he was riding as an observer while Lt. Donald J. Meiers was on radar. The night of November 24, 1944, was partially overcast with a quarter moon. According to a 1945 article in American Legion Magazine about the sightings, Ringwald said, "I wonder what those lights are, over there in the hills," while they were wandering the Rhine Valley near Strasbourg on the French-German border.

They glowed a bright orange, eight or ten of them in a row. Schlueter then saw them from his right wing. When they used Allied ground radar to verify, nothing showed up. Schlueter turns the aircraft to battle, believing the lights to be some kind of German air weapon, only to have the lights disappear.

The males were first silent out of concern that they would be shunned. However, the unit began to hear about the sightings.

While flying at around 800 feet on December 17, 1944, a pilot noticed "five or six flashing red and green lights in 'T' shape" near Breisach, Germany. He felt as if the lights followed him, getting closer "to about 8 o'clock and 1,000 ft." until they mysteriously vanished.

Two additional aircraft crews saw lights on December 22nd. A crew near Hagenau reported seeing two lights in a bright orange glow that seemed to rise 10,000 feet from the ground and follow the aircraft "for about two minutes." The lights "peel off and turn away, fly along level for a few minutes and then go out" after that. Strange Company: Military Encounters with UFOs in World War II by Keith Chester claims that they always seem to be completely under control.

Then there was the incident of Lt. Samuel A. Krasney, who saw a red-glowing cigar-shaped object without wings a few yards from the plane's wingtip. The pilot was told to try evasive manoeuvres by Lt. Krasney, who was understandably alarmed, but the light object remained adjacent to the plane for a few minutes until it "flew off and disappeared."

Ultimately, the pilots gave the lights the nickname "foo fighters," drawing inspiration from the comic strip "Smokey Stover," in which the fireman Smokey would often say, "Where there's foo, there's fire."

After the sightings of foo-fighters were reported by an Associated Press reporter on January 1, 1945, a number of ideas about their origins rapidly surfaced: Flares, weather balloons, or St. Elmo's Fire—a phenomena in which light emerges on the ends of objects during severe weather—were among the items seen. However, none of those hypotheses were accepted by the 415th's members. These things could track aircraft better than flares or weather balloons since they had seen the St. Elmo fire and could tell the two apart.

The Official Government Acknowledgement of Meetings with Alien Visitors

Scepticism And Silence At The Start Of Government UFO Investigations

DO WE LIVE ALONE IN THE COSMOS? If not, are our extraterrestrial neighbours paying us a visit? These kinds of queries, which are usually discussed after a delicious dinner when cigars and port are distributed and philosophical topics take centre stage, have resurfaced on the agenda of the European Parliament.

Musumeci said, "Sightings...have been reported in all corners of the globe for a long time now." Certain phenomena...not only lead to a great deal of confusion, but they also feed speculations about the possible presence of aliens, hypotheses that have no basis in reality as of yet.

Since these events "do not fall within the scope of individual member state competence," the Italian MEP emphasised that the European Commission should give particular attention to the "serious study" being conducted by several European space centres and research facilities.

UFO enthusiasts have maintained for decades that Area 51, a portion of an Air Force installation in southern Nevada, is where the government stores crashed UFOs and potentially aliens. In other UFO news, after a wave of odd sightings by Navy pilots, the Navy recently released new guidelines for troops to report "unexplained aerial phenomena".

The percentage of people who think the government is hiding knowledge concerning UFOs, which is 68% today, is similar to the 71% who said the same in 1996. In terms of age, education, and party affiliation, the results were comparable in both cases across all major demographic categories.

That said, it seems that only around 50% of those who believe the government is concealing knowledge concerning UFOs also believe it is

The Official Government Acknowledgement of Meetings with Alien Visitors

withholding information regarding extraterrestrial space landings. This conclusion is supported by the data showing that a far less percentage of individuals believe the government is more knowledgeable than it is revealing than do those who have personally seen UFO encounters or give them credibility.

According to 33% of American adults, some UFO sightings throughout the years have really been extraterrestrial spacecraft coming to Earth from distant planets or galaxies.

Sixty percent of respondents are sceptical and believe that every observation of a UFO may be explained by either natural phenomena or human action. The other seven percent are not sure.

Of all Americans, 84% have not seen a UFO, whereas 16% claim to have personally seen something they believed to be one.

The majority of Americans (56%) think that people who witness UFOs are seeing something genuine and not merely dreaming, despite the fact that the

The Official Government Acknowledgement of Meetings with Alien Visitors

majority of Americans are dubious of extraterrestrial visits. This is an increase from 47% in 1996, which may be attributed to increased public knowledge of military experimentation and the widespread use of drones, which some people may mistake for unidentified flying objects.

A little over half of Americans, or 49%, believe that there are people substantially like us somewhere in the cosmos, independent of any potential extraterrestrial landings on Earth. Seventy-five percent of people think that life of some kind exists on other worlds.

Western America Leads the Nation in the Belief That UFOs Are Extraterrestrial Spacecraft

Though they are prevalent across all racial and political categories of society, those who believe in UFOs are more likely to live in the West. This is the location of Area 51, as well as the adjacent town of Roswell, New Mexico, the scene of a 1947 crash of some kind that gave rise to the long-running conspiracy theories surrounding Area 51.

The Official Government Acknowledgement of Meetings with Alien Visitors

In the West, 40% of people think that extraterrestrial visitors are responsible for certain UFO sightings. This contrasts with 27% of people in the Midwest and 32% of those in the East and South.

Additionally, Westerners are somewhat more likely than those in other areas to report having personally observed a UFO (20% in the West vs. 12% in the East and 15%–16% elsewhere).

The new survey also reveals notable disparities in UFO sightings by wealth and education, with Americans from lower socioeconomic backgrounds reporting more beliefs than those from higher socioeconomic backgrounds.

Interestingly, the percentage of Protestant and Catholic Americans who believe in UFOs is comparable to the national average. Nevertheless, Americans who do not identify as religious are more inclined to believe in UFOs; 40% believe that some have been visited by extraterrestrial beings.

Governments from all over the world had to contend with a growing public interest in the phenomenon as UFO sightings became more frequent. Project Sign and Project Grudge were both started by the Air Force in the US in 1947 with the intention of looking into UFO accounts. However, the most significant and well-known official UFO research was Project Blue Book, which was started in 1952.

Project Blue Book researched thousands of UFO reports while posing as a scientific investigation in an effort to provide explanations for each incident. Although they made an effort to minimise the importance of UFOs, several of the instances were never fully investigated, creating a lingering air of mystery. The question of whether governments were hiding important information from the public started to take root as Project Blue Book wrapped up its investigations in 1969 and shut its doors for good.

Our trip into the world of UFOs is about to take a new turn as we emerge from the shadow of history and enter the current era. We shall go further into the moment when the government's approach to unidentified flying objects changed for the better in the upcoming chapters. Humanity was about to experience a seismic shift from denial to acceptance—a realisation that would redefine our position in the universe and reveal long-kept UFO secrets.

The Official Government Acknowledgement of Meetings with Alien Visitors

CHAPTER TWO

FROM CONSPIRACY TO DISCLOSURE

Government responses to UFO reports have changed throughout time.

Many governments were interested in the phenomenon in the years after World War II as a result of the rise in UFO sightings. In particular, the United States initiated a number of research initiatives to examine and determine the nature of these aerial mysteries. Project Sign, started by the American Air Force in 1947, was one of the most important initiatives. The main goal of Project Sign was to assess UFO claims and determine whether any of them would pose a threat to national security.

Cover-ups and Conspiracy Theories.

An unsettling pattern emerges in the headlines every few years. Reports of unidentified flying objects (UFOs/UAPs) make headlines. Then a denial is released by a government agency.

Conspiracy theory is the idea that a select few individuals are working behind closed doors and that their goals are to further the interests of their limited group. There is nothing conspiratorial about the existence of aliens. Conspiracy theories only emerge when individuals think that knowledge regarding aliens is being withheld from the general public by a small number of people, usually government insiders.

The Official Government Acknowledgement of Meetings with Alien Visitors

The United States government usually gets the blame when it comes to UFO conspiracy theories. Furthermore, conspiracy theorists often combine the federal government of the United States into a single entity, despite the fact that millions of individuals are employed by "the government," according to academics.

Governments and politicians across the globe, especially the US administration, have been influenced by conspiracy theories about unidentified flying objects to withhold information that may indicate these objects are either the product of alien technology or are being controlled by extraterrestrial intelligence. It's part of a conspiracy theory that says, among other things, that the government of Earth has been taken over by aliens and that, in spite of official denials, the government is in contact with or working with these visitors from space. Some claim to explicitly include Stanford University in the group of those who have hinted to the concealment of UFO evidence.

The Official Government Acknowledgement of Meetings with Alien Visitors

Immunologist Gary Nolan, U.S. Senator Barry Goldwater, astronauts Gordon Cooper and Edgar Mitchell, Israeli Brigadier General Haim Esched (former head of the Israeli Defence Ministry's space programme), British Admiral Lord Hillnorton (former head of NATO and British Chief of Defence), and U.S. Lieutenant General Roscoe H. Hillenketter (first CIA Director) were among them. Other than their reports and testimonials, they have not provided any evidence to support their claims or statements. The Sceptic Commission asserts that there is little or nonexistent evidence to support their views, despite substantial research on the subject conducted by volunteer scientific bodies. Religious scholars have found that there are many new religious groups among individuals who believe in UFO conspiracy theories, the most well-known of which being Heaven's Gate. Islam's State and Scientology.

The Official Government Acknowledgement of Meetings with Alien
Visitors

Chronology of UFO conspiracy theories

Hoaxes involving the "recovered disc" and the Roswell balloon

Roswell Army Air Field announced the discovery of a "flying saucer" in a news release on July 8, 1947. It was not long before the Army withdrew their statement and admitted that the item that had crashed was a regular weather balloon. It wasn't until the late 1970s that the Roswell incident made a comeback and was included in conspiracy theories. The erroneous "disc" was not limited to the Roswell balloon. What seemed to be a disc found in a Grafton, Wisconsin priest's garden was really a standard circular saw blade. In Shreveport, Louisiana, a more advanced fake disc was found. Situated at Wisconsin's Black River Falls. additionally in Clearwater, Florida. A thirty-inch disc was found in a Hollywood garden, according to news reports on July 9. A thirty-inch disc was found in the yard of a Twin Falls, Idaho, residence, according to media reports on July 11. The Twin Falls disc was declared a fraud throughout the country on July 12. An image of the item has been made public. Underneath a plexiglass dome, the device is said to include cables, electrical coils, and radio tubes. Four adolescents admitted to making the CD, the press said.

Winchell with the Soviets

Walter Winchell, a radio host, said on April 3, 1949, that there was solid proof that the flying saucer was a guided missile that was fired from Russia. The Air Force disputed this finding in response. It has been stated that the Air Force asked the FBI to look into Mr. Winchell's claims, but the request was turned down.

Keyhoe and Air Force UFO knowledge

Donald Kehoe's article "Flying Saucers Are Real" appeared in True magazine on December 26, 1949. The former U.S. Marine Corps major Kehoe said

that there were Air Force insiders who were aware
of the saucer's existence and deduced that it was
probably "interplanetary." An unidentified pilot
who thought the Air Force had destroyed the saucer
was cited in the story about the Mantell UFO
encounter. The description provided by the Force
"seems like a cover-up to me." There were
additional explanations for Chili's dogfight with
Gorman and Whitted's UFO experience. According
to the story, "rocket authority at Wright Base" had
determined that the saucer would fly intergalactic,
citing a purported report from the Air Material
Command. The report mentioned fears about the
kind of public hysteria that purportedly followed
the 1938 War of the Worlds broadcast as a potential
reason for the cover-up. Kehoe surmised that such
sightings had probably been happening for a few
centuries at the very least, citing historical records.
True's article created a lot of buzz. Although these
numbers are never easily confirmed, the original
Project Blue Book director, Captain Edward J.
Ruppelt, wrote that there were "rumours among

The Official Government Acknowledgement of Meetings with Alien
Visitors

magazine publishers that the Don Kehoe article published in True magazine is one of the most widely read and widely discussed magazine articles in history." Kehoe turned this essay into a book, Flying Saucers Are Real (1950), which went on to sell over 500,000 copies in paperback. The Air Force rejected the notion that a "flying saucer" was a secret US technology in March 1950.

Alien bodies and Scully

Based on findings from the relevant experts, writer Frank Scully wrote two pieces for Variety magazine in October and November 1949 about alien life that perished in a flying saucer collision. declared to be back on track. He elaborated on this issue in his 1950 book Behind the Flying Saucers, stating that he had two similar experiences in Arizona and one in New Mexico. The 1948 event concerned a disc with a diameter of over thirty metres (100 feet). Time magazine reported a story of a crashed saucer with a humanoid crew in January 1950, however the story was subsequently repeated with scepticism. It was then discovered that Scully had been duped by 'two seasoned confidence artists'. evolved. According to writings by San Francisco Chronicle writer John Philip Kahn in True magazine in 1952 and 1956, Newton and "Dr. Gee" (real name Leo A. Gebauer) were the oil fraudsters that made up Scully.

The 1950s

Governmental and private investigation effort surged in the 1950s in response to growing allegations of evidence hiding and public deception. In 1950, radar physicist Henry Tizard, the Chief Scientific Advisor to the Ministry of Defence at the time, commissioned research that led to the start of the UK Ministry of Defence's UFO project. His determination that UFO encounters should not be written off without appropriate scientific investigation led the CIA to create the Flying Saucer Working Group, or FSWP. Nicholas Mariana, a baseball manager in Montana, took pictures of many UFOs in August 1950. 16 mm film. Mariana

The Official Government Acknowledgement of Meetings with Alien Visitors

said that crucial material had been cut out of the video after Project Blue Book was contacted and examined it; this was evident when the tape was played again and revealed the item to be a disc. The event was covered by the national media. Working for the Canadian Department of Transportation, Wilbert B. Smith was a radio engineer who was curious in flying saucer propulsion technology and questioned the veracity of the claims made in the recently released Scully and Kehoe book. had pondered. He was in charge of the Canadian Embassy in Washington, D.C., in September 1950. Make arrangements to speak with American representatives in an effort to get the facts straight. Robert Thurbacker, a consultant to the Defence Research and Development Committee and physicist, informed Smith. Additional correspondence on Kehoe's need for authorization to release an additional article on Smith's UFO propulsion idea suggested that Bush and his team were functioning outside the purview of the Research and Development Board. Project Magnet, a modest UFO research project for the Canadian government, was established as a result of Smith's briefings to senior government officials. The late 1970s saw the discovery of Smith's private files and Canadian records. By 1984, more claims had surfaced about the existence of a top-secret UFO monitoring group made up of scientists and military officials, known as the Majestic 12. Vannevar Bush was brought up once again. In the 1980s, Serbacher was also interrogated, and his testimony verified details found in Smith's memos and letters. In public interviews throughout the 1950s and early 1960s, Smith said that he had been sent crashed UFO data for examination by an undisclosed top-secret US government organisation. A few weeks after the Robertson Panel, the Air Force published Rule 200-2, instructing staff at Air Force bases to categorise and disclose all unresolved cases and to publicly discuss UFO investigations only once they are found to be closed. I told you not to get near my eyes. In addition, the newly established Air Defence Force

The Official Government Acknowledgement of Meetings with Alien Visitors

4602nd Air Intelligence Squadron (AISS) started conducting UFO research flights. Only the most significant UFO occurrences affecting intelligence and national security were to be investigated by the 4602nd AISS. They were purposefully taken from Blue Book so that Blue Book could handle the less important reports.

Keyhoe and The Conspiracy of the Flying Saucer

Donald Kehoe published a new book in 1955 in which he severely charged the US government of engaging in a plot to hide its knowledge of flying saucers. According to Kehoe, there is a "silent group" directing this plot. For the next twenty years, Kehoe's belief in this, the book featured claims about the possibility that a "orbiting space station" or "moon base" had been discovered, the discovery of which caused public panic. Folk historian Curtis Peebles says of this, "The flying saucer conspiracy marked a shift in Kehoe's belief system. Flying saucers were no longer a central theme; they now belonged to the Silent Group and its cover-up." possible result The Bermuda Triangle disappearance mythology was also included in the flying saucer theory. Kehoe's sensationalised reports of strange formations on the moon were eventually shown to be optical tricks.

The Philadelphia Experiment and Carl Allen

With his book The Case of UFOs, published in 1955, Morris K. Jessup rose to considerable prominence by putting up the argument that UFOs were a mystical topic deserving of further research. In January 1956 Jessup started receiving a series of letters from "Carlos Miguel Allende" (later identified as Carl Meredith Allen), who he suspected were UFOs "probes of a'solid' and 'vague' nature." Jessup also "linked ancient sites with prehistoric parascience." The navy vessel was successfully rendered invisible by Allende, who is said to have told Jessup not to explore the floating UFO. However, the warship was mysteriously transferred from Philadelphia to Norfolk, Virginia, and back again. We have crafted a story about risky

The Official Government Acknowledgement of Meetings with Alien Visitors

experimenting. It was believed that the crew of the ship had a number of adverse affects, such as being "frozen" in place, numb, and insane. Author of ghost stories Charles Burlitz began promoting the Philadelphia Experiment in 1975, and a fake film based on the tale was released in 1984. When Jessup visited the Office of Naval Research in 1957:67, he was handed a copy of his book with annotations. Handwritten notes in three distinct hues of blue ink were found in the book's margins; they seemed to record a conversation between the three. one person. They spoke about concepts for extraterrestrial propulsion and flying saucers, and they were worried that Jessup was coming too near to finding out about their technology. : 27–29, 35, 65, 80, 102, 115, 163–165 Jessup said the remark mirrored what he had read in Allen's letter. :9 Jessup's book with Allen's scrawled notes outlasted it; twelve years later, Allen said he penned all the annotations "to frighten Jessup to the brim." When Garland, Texas-based Varo Manufacturing Corporation, under contract with ONR, started making mimeographs of the book with Allen's comments and letters to Jessup, it became unique. :9 These replicas were dubbed "Varo editions". :6 Numerous follow-up "Philadelphia Experiment" books, documentaries, and movies centred on this. Karl Allen has provided evasive or nonexistent answers to several journalists and researchers who have attempted to contact him for further information throughout the years.

The Report on Unidentified Flying Objects and Edward Ruppelt

'Men in Black' and Grey Barker

Grey Barker released I Knew Too Much About Flying Saucers in 1956. The book advocated the theory that witnesses to UFO sightings would see men in black and be warned to remain silent. The guys in black are said to be government operatives who intimidate and harass witnesses to unidentified flying objects. In the Sceptical Inquirer article "Grey Barker: Mythmaking Friend, Grey Barker," Barker discusses the Men in Black and how government organisations attempted to stifle public curiosity about unidentified flying objects. 's essays from the 1950s could have included "a piece of truth." However, Barker is thought to have greatly exaggerated the truth. Sherwood disclosed in the same Sceptical Inquirer piece that he collaborated with Barker to produce a brief, fake trailer that made references to Men in Black in the late 1960s. Originally appearing as truth in some of Barker's and Raymond A. Palmer's periodicals, this notification was published. own printed works. In the narrative, Sherwood (author of "Dr. Richard H. Pratt") said that after discovering that UFOs were time-traveling spacecraft, "negroes" gave him the command to be quiet. Afterwards, Barker wrote, "Clearly the fans swallowed this whole thing," to Sherwood.

The 1960s

Atmospheric scientist James E. MacDonald made the suggestion in articles, seminars, and letters that

The Official Government Acknowledgement of Meetings with Alien Visitors

the US government was mismanaging the evidence for the alien theory during most of the 1960s.

Pentacle Memorandum" and Jacques Vallee

File organisation was assigned to researcher Jacques Vallee in June 1967 after Project Blue Book researcher J. Hynek Allen. A message dated January 9, 1953, written to an Air Force officer Edward J. Rappelt's assistant and assigned to Blue Book was discovered by Vallée among those papers. ..The letter asked for agreement on "what can and can't be done" and referenced to an examination of thousands of hitherto unidentified UFO sightings. It was signed "H.C. Cross," but Mr. Vallee decided to refer to the author as "Pentacle." It ought to be covered in the 1953 Roberson Panel discussion. The memo's release "would cause even more uproar among foreign scientists than among Americans. It will prove his evil nature," Vallee wrote in his 1967 journal. UFOs are not real. "

The Official Government Acknowledgement of Meetings with Alien Visitors

The landing at Holloman Air Force Base and the Emenegger documentary

According to Clark, the encounter from 1973 may have been the first sign that the US government was working with ET. In order to make a documentary, Robert Emenegger and Alan Sandler from Los Angeles, California got in touch with Norton Air Force Base personnel that year. According to Emenegger and Sandler, Paul Shattle and other Air Force personnel have made the decision to include UFO material in the documentary. One of the features will be actual video of a UFO landing at Holloman Air Force Base in New Mexico in 1971. reported to have made a proposal. Emenegher also claimed that on a visit of Holloman Air Force Base, he saw authorities having meetings with extraterrestrials. This won't likely be the first time the has seen these aliens, according to Mr. Emenegger's U.S. military sources, who were "monitoring signals from an alien group unfamiliar to them. And did their ET guests know anything about them?" UFOs: Past, Present, Future, narrated by Rod Serling, was the title of the documentary when it was aired in 1974. Just a little segment of Holloman UFO film is shown; the majority of the landing is depicted and reenacted. Shattle claimed to have seen the aforementioned movie many times and that it was genuine in 1988. Rod Serling, Burgess Meredith, and Jose Ferrer broadcast Robert Emenegger's documentary report "UFO: It Began" on television in 1976. A few scenes were reenacted using eyewitness accounts and the findings of both military and civilian government investigations. The setting of this documentary is a fictitious UFO landing at Holloman Air Force Base. The "startling" similarity between Emenegger's 1973 portrayal of the landing on Holloman and Steven Spielberg's 1977 portrayal of the landing on Devil's Tower in Close Encounters of the Third Kind is well recognised. are recognised. UFO researcher Richard Dolan talks about the Emenegger documentary in the 2013 film Mirage Men, saying, "As many have questioned, the video was some type

The Official Government Acknowledgement of Meetings with Alien Visitors

of 'disclosure. I have been wondering whether it was a failed effort at

"Cosmic Watergate" and J. Allen Hynek

American astronomer J. Allen Hynek worked for the Air Force on three projects in the United States: Project Sign (1947–1949), Project Grudge (1949–1951), and Project Blue Book (1952–1969). Hynek was mocked for disproving his most well-known fallacy, which was to propose that the mass sightings over Michigan could have been caused by "swamp gas." By 1974, sceptics had publicly blasted Blue Book as "the watergate of the universe," claiming that 20% of Blue Book incidents remain unexplained. Other

Option 3 and a covert space initiative

The fake documentary Alternative 3, which aired on British television on June 20, 1977 (really on April Fool's Day), spawned a number of tales that eventually became popular UFO conspiracy theories, according to Jerome Clark. The paperback version of the documentary was later published. We can return," he said. According to a made-up research given in this episode, the lost scientists were part of a covert US/Soviet space programme long before interplanetary space flight was widely acknowledged. a fake Apollo astronaut claimed to have come upon a mystery lunar outpost while walking on the moon, implying that it was plausible. It was claimed that scientists had concluded that long-term life on Earth's surface was impossible due to pollution, which would result in catastrophic climate change. Richard Marner's portrayal of physicist "Dr. Karl Gerstein" said that he had offered three solutions to the issue in 1957. The first was a sharp decline in the number of people on Earth. Building massive subterranean shelters to hold public servants and some citizens while the environment stabilised was a backup plan. Using a way station on the Moon, a third option—dubbed "Alternative 3"—was to settle on Mars. The last scenes of the movie show creatures being seen on Mars' surface.

The Official Government Acknowledgement of Meetings with Alien Visitors

Paul Benennewitz

In the late 1970s, Paul Bennewitz of Albuquerque, New Mexico, became the centre of controversy.

Conspiracy theories around Jesse Marcel and Roswell

UFOlogist Stanton Friedman spoke with Jesse Marcel in February 1978. From the site of the Roswell wreckage recovery to Fort Worth, where reporters reportedly saw items said to be a portion of the recovered object, he is the only person known to have accompanied him. Marcel made claims to the press in 1947 that were in conflict with his assertions. Marcel's first recorded interview was included in the co-written documentary "UFO's Are Real" in November 1979. After a restricted theatrical run, the movie was syndicated on television. Marcel's tale gained widespread notice when it was published in the sensational tabloid National Enquirer on February 28, 1980. September 20, 1980 saw the premiere of the TV show In Search of... aired an interview wherein Marcel elaborated on his involvement in the 1947 press conference, saying, "They wanted me to comment, but I didn't have that freedom. So all I could do was keep my mouth shut. And it was General Ramey who argued - the newspaper, the newsman. It was and should have been forgotten." Just a weather balloon, that's all. Of course, we both had different perceptions." He gave his final interview on HBO's America Undercover. denied the existence of the body. From 1978 to the early 1990s, UFO researchers such as Stanton T. Friedman, William Moore, Karl T. Floch, and the team of Kevin D. Randle and Donald R. Schmidt were associated with UFOs. I interviewed dozens of people who claimed that In the 1990s, the U.S. military released two reports that revealed the identities of downed planes. It was a surveillance balloon by Project Mogul. Nonetheless, the Roswell incident continues to attract media attention, and conspiracy theories surrounding it persist. Roswell has been described as "the most famous, most thoroughly investigated, and most thoroughly debunked UFO claim in the world."

Cooper, Gordon

After the incident was reported in the Los Angeles Times and other newspapers, the official explanation was that photographers had captured a weather balloon distorted by the hot desert air. However, astronaut Gordon Cooper claimed in 1981 that high-definition footage of the flying saucer, taken by two ranging cameramen at Edwards Air Force Base on May 3, 1957, had been suppressed. Cooper saw a developed negative of the object and said it clearly showed a dish-shaped object with a dome. Things like holes or ports in the top and dome. Cooper and another witness later confirmed the story when interviewed by James McDonald.

Magnificent 12

The term "extraterrestrial life" (or EBE) was used in the so-called Majestic 12 documents, which first surfaced in 1982 and suggested that there was a high-level secret US government interest in UFOs dating back to the 1940s. The FBI declared the documents to be "totally fake," and many ufologists believe they are elaborate hoaxes.

Cattle mutilations and Linda Moulton Howe

Linda Moulton Howe is an enthusiast of conspiracy theories, such as the idea that the U.S. government is working with aliens, and the idea that the Cattle Mutations are alien in origin.

Milton William Cooper and George C. Andrews

Conspiracy theorist George C. Andrews accused the CIA of being involved in the Kennedy murder in his 1986 book Extraterrestrials Among Us. Michael Birken, a researcher on extremism, argues, "Andrew's political views are almost indistinguishable from those associated with militias, but the fact that he places extraterrestrials at the top of the conspiracy alone is enough to make him an expert in UFO research." It's recognised as a home," he said. For Burkun, "The publication of 'Extraterrestrials' marked the beginning of a frenetic period of UFO conspiracy theories from 1986 to 1989." Pale Horse informed Majestic 12 that Kennedy was "going to reveal to the American people the existence of aliens." Pale Horse subsequently claimed to have been assassinated. Bill Cooper, the idea's originator, released the well-known conspiracy theory book Behold. The notion of Andrews and Cooper contributed to "a conspiratorial form of UFO speculation that Jerome.

Bob Lazar and Area 51

Following Delta Force training for Operation Eagle Claw, an unsuccessful American hostage rescue mission in Iran, the name "Area 51" was coined in the popular press in 1980. The location was discussed in the media once again in 1984, after the government's acquisition of the nearby acreage. In a special interview with investigative reporter George Knapp on the Las Vegas television station KLAS in

November 1989, Bob Lazar discussed his purported work with S-4. In an interview with Knapp, Lazar said that at first he believed the saucer to be a covert ground-based vehicle and that many UFO sightings were likely the result of its test flights. Lazar examined the disc carefully and was given many papers that explained its origins, leading him to the conclusion that it had to be alien in nature. He asserts that the "grey" aliens are from the Zeta Reticule system and that they are warping space using moscovium, an element that disintegrates quickly. The Los Angeles Times claims that he never got the degrees from Caltech and MIT that he claimed to hold. By 1991, news organisations in Nevada were covering visitors who were heading to the Groom Lake region in an attempt to see a UFO.

Cover-Up of UFOs?: Present!
Actor Mike Farrell starred in the two-hour television spectacular "U.S. Cover-Up: Live!" UFO on October 14, 1988.at

The Official Government Acknowledgement of Meetings with Alien Visitors

1990s = Since at least the middle of the 1990s, a number of conspiracy theories have been making the rounds on the Internet, and these beliefs are supported by a set of papers known as the Blanton Files. The most common person to credit them is Bruce Alan Walton, who says he was abducted by aliens and spoke with people "living inside the Earth" via a "altered state of consciousness." According to the description, these files are "high fantasy" with "complex and intricate plots."

Phil Schneider and Dulce Base
A guy going by the name Philip Schneider basically supported an updated version of the aforementioned idea in 1995 while attending many UFO gatherings. Schneider identified himself as the offspring of a U-boat captain who defected after being taken prisoner by the Allies. Schneider claims that his dad took part in the Philadelphia Experiment. Schneider said that he was exposed to a variety of secret material and his own experience with the EBE as a consequence of his involvement

in the building of Deep Underground Military Bases (DUMBs) around the United States. He made the decision to share his tale after claiming to have escaped the catastrophe at Dulce Station. Schneider died on January 17, 1996, according to tradition, however some of his supporters said he had been killed.

In the 2000s, Chuck Misler and Mark Eastman released Alien Encounters (ISBN 1-57821-205-7) in 2003. The ideas discussed above, particularly Cooper's, are essentially repeated and presented as truth in this work.

MoD classified documents

The UK Ministry of Defence originally posted eight files pertaining to UFO encounters from 1978 to 1987 on the National Archives website on May 14, 2008. There were 200 files that needed to be published by 2012. These files include letters that members of the public sent to Margaret Thatcher and other government officials as well as the

Department of Defence. You may download information. Up until the base's closure in the 1980s, public relations officers at U.S. bases regularly distributed copies of Lieutenant Colonel Halt's letter to the British Ministry of Defence (see above) describing the sighting at RAF Woodbridge, without providing any more commentary. was released in). In response to a Freedom of Information Act request, the Department of Defence disclosed the papers. An Englishman's report of a "light in the sky" was in the file.

These gaps in the conspiracy theory did not, however, make the road to official government recognition of UFOs any easier to travel. Governments were reluctant to change their official stance due to potential public discontent or a reduction in confidence in government agencies; however, the emergence of whistleblowers and the publication of documents marked a change in the public's perception of UFOs and set the stage for subsequent events that would upend the paradigm

The Official Government Acknowledgement of Meetings with Alien Visitors

of denial and secrecy surrounding unexplained flying objects.

Disclosure

Within the UFO conspiracy scene, the idea of "disclosure" gained traction in the early 2000s. Activist lobbying organisations pushed for complete disclosure, arguing that the government had concealed and classified evidence about extraterrestrial contact.

To further the idea, Steven M. Greer established the Disclosure Project in 1993. At a news conference in May 2001, Greer requested that Congress undertake hearings on "secret U.S. involvement with UFOs and extraterrestrials" at the National news Club in Washington, D.C. "The strangest ever news conference hosted by Washington's august National Press Club," was how one of the attending BBC reporters put it. Sceptics and representatives of the US Air Force ridiculed the assertions made by the Disclosure Project.

The Official Government Acknowledgement of Meetings with Alien Visitors

A "Citizen Hearing on Disclosure" was conducted in 2013 by the production firm CHD2, LLC, and it took place at the National Press Club in Washington, D.C. from April 29 to May 3. The organisation provided $20,000 to each of the following former lawmakers: Carolyn Cheeks Kilpatrick, Roscoe Bartlett, Merrill Cook, Darlene Hooley, Lynn Woolsey, and former U.S. Senator Mike Gravel. In exchange, they presided over panels of scholars and former military and government officials debating UFOs and extraterrestrials.

Other similar organisations include the 1977-founded Citizens Against UFO Secrecy.

After its first results showed that a significant percentage of UFO encounters remained unidentified, Project Sign concluded that further investigation was required. This revelation led to the creation of Project Grudge, which aimed to refute and stop UFO reports. Grudge's attitude, which was usually disdainful and disregarded the most compelling UFO encounter tales, increased the public's distrust

CHAPTER THREE

BREAKING SILENCE: THE ROAD TO OFFICIAL RECOGNITION

The mid-20th century saw a turning point in the history of UFOs and government participation as a rise in sightings and public curiosity led to the establishment of official channels for the investigation of these enigmatic phenomenon. This chapter examines how governmental organisations gradually changed their stance on the UFO mystery from one of blatant denial to one of greater moderation. We will explore the turning points in this shift, the major

The Official Government Acknowledgement of Meetings with Alien Visitors

participants, and the debates that accompanied their attempts.

The Turning Point: Important UFO Events and Their Consequences

In the history of UFOs, specific events had a crucial role in prompting government authorities to take the phenomenon more seriously. The report that pilot Kenneth Arnold saw nine crescent-shaped objects flying at great speeds over Mount Rainier in Washington State in June 1947 is among the most well-known incidents. Arnold's story attracted media attention and sparked widespread interest, which sparked more

The Official Government Acknowledgement of Meetings with Alien Visitors

UFO reports and attracted the attention of the U.S. military.

According to Mark Rodeghier, scientific director of the Centre for UFO Studies in Chicago, "the effort to detect, track and measure the UFO phenomenon in the field, in real time, has recently entered a new phase." Space.com reported this information. "Software tools have advanced, technology has advanced, and the present fascination in UFOs has drawn in new, skilled specialists.

"As a consequence, we will have even more evidence — as if it was needed — that the UFO phenomenon is real and can be studied scientifically," he said.

The Official Government Acknowledgement of Meetings with Alien Visitors

Regarding UFOs, the US government is likewise becoming more serious. In 2021, Congress called for the creation of an official agency to conduct a "coordinated effort" in gathering and analysing data pertaining to UAPs, in response to a study on UAPs released by the U.S. military and intelligence community.

While the office will not specifically focus on the search for extraterrestrial life, President Joe Biden signed the $768.2 billion National Defence Authorization Act for Fiscal Year 2022 on December 27, 2021. The act included a provision for the creation of the Airborne Object Identification and Management Synchronisation Group.

The Official Government Acknowledgement of Meetings with Alien Visitors

The Development of Project Blue Book and Its Disputations

In 1952, the US Air Force launched Project Blue Book in response to the public's rising unease and interest in UFOs. The project's goal was to collect, investigate, and analyse UFO reports to see whether they were a risk to national security.

The United States Air Force's methodical investigation into unexplained flying objects, known as Project Blue Book, ran from March 1952 until its end on December 17, 1969. Captain Edward J. Ruppelt served as the project's first director. It was based at Wright-Patterson Air Force Base in Ohio and followed comparable efforts like Project Sign, which was started in 1947, and Project Grudge, which was started in 1948. Determining if UFOs posed a danger to national security and conducting a scientific analysis of UFO-related material were the two main objectives of Project Blue Book.

Numerous thousands of UFO reports were gathered, examined, and recorded. Project Blue Book was cancelled in 1969 as a consequence of the Condon Report, which said that it was doubtful that studying UFOs would result in significant scientific advancements, and the National Academy of Sciences' assessment of the report.

After collecting 12,618 UFO claims by the time Project Blue Book came to an end, it was determined that the majority of them were misidentifications of conventional aircraft or natural phenomena like clouds, stars, etc. The National Reconnaissance Office claims that flights of the formerly-secret U-2 and A- reconnaissance aircraft might account for a lot of the reports.12.701 complaints were deemed inexplicable after thorough investigation. The Freedom of

Information Act permits access to the archived UFO reports, however all witness names and other private information has been suppressed.

Project Blue Book's critics said that, despite its claims to the contrary, it was little more than a public relations blitz to quell UFO sightings and satisfy public interest while withholding classified material.

As the experiment progressed, the findings shifted dramatically, leaving some instances labelled as unidentified and allowing speculation to grow. However, it quickly became apparent that the government was reluctant to release any data that may be construed as proving the existence of alien life or extremely advanced technology that is inaccessible to humans.

Governmental Agencies Under Stress: The Need for Openness

As time went on, the UFO phenomena gained more and more attention, and people, academics, and organisations advocating for more openness in UFO investigations increasingly criticised government agencies. The increasing number of sightings and the public's need for answers have led to proposals for open hearings on the subject.

In 1968, the House Committee on Science and Astronautics conducted a hearing on UFOs during which scientists, military officers, and citizen witnesses testified. Even

The Official Government Acknowledgement of Meetings with Alien Visitors

while the committee's final report did not provide definitive evidence of alien origins for UFOs, it did recognise the need for additional scientific inquiry and advocated for a more systematic approach to data collection.

The government was compelled to reevaluate its approach due to the committee's actions and the diligent efforts of UFO researchers. The public's insatiable hunger for UFO knowledge led some politicians to consider treating UFO sightings in a more transparent manner.

The beginning of the 1970s saw the interesting development of certain UFO-related materials becoming declassified. Documents that had been kept hidden before were made public thanks to FOIA requests and the tenacity of researchers. While most of these documents did not provide conclusive evidence of alien life, they did demonstrate the government's interest in UFOs and the lengths it went to investigate and classify individual cases.

The sudden openness was a significant improvement, even if it was only temporary, since it indicated a shift in the government's approach to the UFO phenomenon. It demonstrated a willingness to accept the public's curiosity in UFOs and the growing realisation that more research was necessary than just denial.

The Official Government Acknowledgement of Meetings with Alien Visitors

CHAPTER FOUR

SECRET UFO STUDIES: BEHIND CLOSED DOORS

As UFO sightings and interactions that defied conventional explanations became public, governments everywhere launched covert inquiries and research operations. Research into the enigma of unexplained flying objects was kept in the darkest recesses of clandestine study. In this part, we go into the murky realm of top-secret government UFO programmes and explore the motivations behind their inception.

Many military and intelligence agencies have established dedicated teams to look into UFO claims. These organisations, which ranged from the US Defence Intelligence Agency (DIA) to the Soviet Union's KGB, closely monitored the occurrence.

The Official Government Acknowledgement of Meetings with Alien Visitors

The two major objectives were to evaluate the possible danger presented by UFOs and to make use of any technical advancements that may result from studying these alien objects.

One of the most well-known secret UFO initiatives is the United States' Advanced Aerospace Threat Identification Programme (AATIP). In order to investigate UFO sightings reported by both military personnel and civilians, the Pentagon launched AATIP in 2007. The program's existence remained unknown to the public until its exposure in 2017. As we dig further into AATIP's activities, we will look at its outcomes, challenges, and the controversies surrounding its demise.

The Official Government Acknowledgement of Meetings with Alien Visitors

Unreleased Secret Documents: An Examination of Government UFO Research

In the group of real believers in Unidentified Flying Objects, there is one issue that keeps coming up: If the government has nothing to hide, why is it hiding so many UFO records?

The head of the Federation of American Scientists' Project on Government Secrecy in Washington, D.C., Steven Aftergood, said, "Well, it turns out that the government does have something to hide, but it has nothing to do with extraterrestrials."

A document with the codename "Top Secret Umbra," which designates the highest and most sensitive level of communications intelligence, has come to light.

Originally submitted in a 1980 lawsuit by the National Security Agency (NSA) to support the withholding of UFO documents, the once-classified declaration was used as evidence. The paper is mostly declassified, while certain portions have been removed, purportedly to safeguard employee identities and prevent exposure of NSA techniques, expertise, and overseas contacts.

In response to a Freedom of Information Act (FOIA) request from researcher Michael Ravnitzky, the document titled "In Camera Affidavit of Eugene F. Yeates: Citizens Against UFO Secrecy v. National Security Agency, October 9, 1980" was released on November 3 in redacted form and posted on the Federation of American Scientists website.

outside signals
Reading the text will help one understand how the NSA, a super-secret organisation, became involved in the UFO scene.

The Official Government Acknowledgement of Meetings with Alien Visitors

America's cryptologic agency, the National Security Agency/Central Security Service, was established in November 1952. It gathers intelligence on foreign signals and transmits it while coordinating, supervising, and carrying out very specialised tasks to safeguard federal information networks in the United States.

The NSA is a state-of-the-art location for data processing and communications since it is a high-tech organisation. It serves as a government hub for research and analysis on foreign languages.

According to the recently revealed 1980 document, the NSA has 239 UFO-related materials in its possession, 79 of which came from other government organisations. An NSA official's report from a UFO symposium is included in one of the documents. These reports were written mostly between 1958 and 1979.

The Official Government Acknowledgement of Meetings with Alien Visitors

falsified data

The disclosed document has many fascinating names for UFO materials tied to the NSA, including Survival Questions and UFO Hypothesis.

UFOs and the Intelligence Community Blind Spot to Surprise or Deceptive Data is another term that is mentioned. An anonymous NSA employee is said to have identified "a serious shortcoming" in the agency's communications intelligence (COMINT) reporting and interception processes in this seven-page, undated, unauthorised draught of a monograph. In other words, "the incapacity to appropriately react to unexpected information or purposefully misleading data."

The unnamed author cites the UFO phenomena to support his contention that the United States' intelligence community's incapacity to handle this kind of unexpected material has a negative impact on the country's ability to collect information.

The Official Government Acknowledgement of Meetings with Alien Visitors

The Role of Military and Intelligence Agencies in UFO Research

The first wave of UFO sightings coincided with the start of the Cold War in 1947 between the US and the USSR. When Kenneth Arnold reported seeing nine disk-shaped objects in 1947 close to Mount Rainier, it started a countrywide chain of sightings that resulted in reports from both military and civilian sources. Air Force Gen. Nathan Twining launched Project SIGN in 1948, which subsequently became Project SAUCER, to gather and assess UFO evidence within the government in response to the possible national security threat.

The Air Force came to the conclusion that while UFOs were genuine, they were often explicable as Project SIGN developed under the Air Material

The Official Government Acknowledgement of Meetings with Alien Visitors

Command in Dayton, Ohio. They attributed sightings to public panic, hoaxes, or misinterpretations. However, the likelihood of alien occurrences persisted, prompting the suggestion that military intelligence should maintain command over inquiries. UFO sightings continued after Project SIGN was terminated in 1949, leading to the establishment of Project BLUE BOOK in 1952 and its continued prominence as a primary research project for UFOs in the 1950s and 1960s.

From 1947 to 1952, as tensions during the Cold War and the Korean War increased, the CIA kept a careful eye on the Air Force's UFO studies. A wave of sightings in 1952 prompted the CIA to organise a special research committee, acknowledging the distant possibility of interplanetary planes and raising concerns about possible security dangers. The Truman government were alarmed in July 1952 when a series of unexplained blips on radar scopes accumulated, but they provided an explanation to allay public fears. Although it was kept a secret

The Official Government Acknowledgement of Meetings with Alien Visitors

from the public, the CIA's interest in UFOs had a significant impact on how the U.S. government would officially view them for many years to come.

When the CIA Study Group investigated UFO claims in 1952, during the height of the Cold War, they noticed a striking lack of Soviet references, which made them assume intentional government suppression. The group was concerned about the possibility of flooding the US air warning system with UFO sightings, giving the Soviets an edge in a nuclear assault, since they feared that the Soviet Union would employ UFOs as psychological weapons.

The CIA Study Group considered the flying saucer phenomena to be a major national security threat because to the growing tensions of the Cold War and the increased capabilities of the Soviet Union. They imagined that the Soviet Union would take use of UFO claims to create widespread panic in the US and maybe overwhelm the air warning system.

H. Marshall Chadwell, Assistant Director of OSI, felt the issue was important enough to need the National Security Council's attention.

In December 1952, Chadwell briefed DCI Smith, which led to immediate action. A document was written and sent to the National Security Council suggesting that the study of UFOs be given top attention. In addition, Chadwell called for the creation of an independent study investigating the UFO issue under the direction of highly qualified experts. Smith oversaw the creation of an intelligence directive for the National Security Council after receiving this briefing.

After the Intelligence Advisory Committee had a meeting in December 1952 to study UFOs, the Robertson Panel was established in January 1953. The team, which included of eminent nonmilitary experts such as physicist H. P. Robertson, examined UFO data and came to the conclusion that the majority of sightings had plausible causes. All of

The Official Government Acknowledgement of Meetings with Alien Visitors

them concluded that there was no proof of alien encounters or a direct danger to the country's security. In order to stop widespread panic, the group suggested disproving UFO accounts and suggested public education initiatives.

The CIA became less interested in UFOs after the Robertson Panel. The Physics and Electronic Division of OSI took over as the UFO monitoring division, suggesting a lessened focus on the problem. Concerns remained despite some authorities calling for the project to be abandoned since there was no fresh information available, especially in light of accounts of UFOs seen abroad and Soviet developments in unorthodox aircraft. The CIA's interest in UFOs began to decline in the 1950s, but the threat of future technical advancements by the Soviet Union kept the subject on the intelligence radar. Owing to its distinct flying characteristics, the CIA's U-2 project, a high-altitude surveillance aircraft, caused a spike in UFO reports in the mid-1950s. Unaware of the top-secret experiment, pilots and air traffic controllers reported seeing strange objects in the sky.

Because of its silver sheen, reflecting surfaces, and high altitude flights at dawn and dusk, the U-2 often gave the impression to those below that it was fiery and unidentifiable. Although they were aware of the U-2 flights, the Air Force BLUE BOOK investigators kept the real nature of the operation a secret while attributing numerous UFO reports to natural causes. According to CIA estimates, human reconnaissance aircraft, especially the U-2, were responsible for more than half of the UFO claims that occurred in the late 1950s and early 1960s. Conspiracy theories were fueled by the false public pronouncements that resulted from this. Unexplained UFO sightings decreased from 5.9% in 1955 to 4% in 1956.

In the 1950s, the Robertson panel study on UFOs came under examination amid growing calls for

The Official Government Acknowledgement of Meetings with Alien Visitors

openness. Its publication was desired by civilian UFO organisations, who were fuelled by UFOlogists and former Air Force officers. The Air Force requested declassification authorization from the CIA, but they refused, resulting in a sanitised version that did not reveal the CIA's role.

As UFO concerns continued throughout the 1960s, a new committee headed by Dr. Edward Condon carried out a thorough investigation. The committee's 1969 report, which was met with scepticism from the public, suggested that the Air Force discontinue its UFO study programme. UFO conspiracy ideas persisted even after Project Blue Book was terminated. Suspecting that the CIA was hiding UFOs, Ground Saucer Watch (GSW) filed a FOIA case against the agency in the late 1970s. Citing national security, the Agency disclosed hundreds of documents but kept others hidden. Despite not demonstrating a substantial UFO participation, the published papers stoked dramatic media attention and widespread scepticism.

The CIA kept a low profile on UFOs throughout the 1980s, concentrating on parapsychology and counterintelligence elements of the phenomenon. There were continued allegations of document concealment, particularly those pertaining to the 1947 Roswell event. Later, controversial Majestic-12 papers that had been rumoured to suggest a top-secret UFO committee were shown to be false. Like JFK murder conspiracy theories, the UFO phenomenon is resistant to normal scientific analysis because of its emotive appeal and general mistrust of government operations. The public's continued interest with the UFO enigma is ensured by the persistence of the belief in alien life.

The Official Government Acknowledgement of Meetings with Alien Visitors

CHAPTER FIVE

THE LATEST UFO HEARD: A CONGRESSIONAL STUDY

In the highly anticipated UFO hearing before Congress on Wednesday, July 27, 2023, three veterans of the armed forces gave powerful and persuasive testimony. Standding among them was a former Air Force intelligence officer, steadfast in his assertion that the United States government had been working covertly on an extensive "multi-decade" operation to reverse engineer returning vessels. It was surprising to hear him say later that non-human "biologics" had been successfully recovered by the US from alleged crash sites.

The Official Government Acknowledgement of Meetings with Alien Visitors

Although "little green men" were discussed, the conversation was primarily focused on how urgently procedures for reporting unidentified aerial phenomena (UAPs) needed to be improved. These sightings can occur in both air and water, and they are increasingly referred to as "anomalous" phenomena rather than "aerial" phenomena. There have been resolute demands to stop stigmatising pilots who report UAP sightings. There have also been repeated requests for government efforts entrusted with solving these paranormal riddles to be strictly regulated. Sincerity, transparency, and accountability are now necessary!

"Retired Maj. David Grusch, a courageous whistleblower who appeared before the national security subcommittee of the House Oversight Committee and revealed crucial facts, now stands out as a vital member of the Pentagon's UAP Task Force. While being prohibited from taking part in many government UFO projects, Grusch has

The Official Government Acknowledgement of Meetings with Alien Visitors

undeniable information of the precise locations of UAPs in U.S. possession!

Because of relentless political pressure and the public's need for answers, federal and military institutions were compelled to provide a plethora of material about unusual aircraft encounters. All the same, it is important to realise that many of the many sightings have been connected to everyday sources, such weather balloons, drones, flying debris, and ordinary birds.

But Defence Department spokesperson Susan Gough has adamantly said that despite the Pentagon's persistent efforts, no credible proof has been uncovered to back up claims of extraterrestrial element possession or reverse engineering, either in the past or in the present.

Grusch, on the other hand, audaciously claims that the US has really acquired "non-human" biological material from the drivers of these shadowy cars.

The Official Government Acknowledgement of Meetings with Alien Visitors

Supporters of the UAP project who have firsthand experience of it and those who continue to be actively engaged attest to this. Aware that such disclosures can have unintended consequences, Grusch audaciously pledges to disclose the realities that have been withheld from the public while keeping further details hidden.

He highlights that those with impeccable integrity and a lengthy history of serving our nation provided the data that supported his statement. Over the course of four years, while working on the UAP task force, Grusch painstakingly collated the testimony of over forty witnesses, which provided the foundation for his compelling arguments.

The importance of this clandestine evidence has been purposely concealed from Congress, as Grusch reveals. Grusch keeps legislators at bay throughout the discussion by claiming that he can only provide further information in a SCIF, or sensitive compartmented information facility. This

The Official Government Acknowledgement of Meetings with Alien Visitors

restriction stands even in the face of concerns about possible firsthand encounter with extraterrestrial life or the malicious withholding of information concerning "extraterrestrial technology" and potential victims. The mystique surrounding this enigmatic place is revealed by his silence.

Moreover, Grusch courageously reveals that since coming out, he and his colleagues have been the victims of administrative terrorism and deliberate prosecution, which has periodically made him fear for his life. It is evident from the brutality he experienced in both his personal and professional life that there are dangers associated with revealing these long-suppressed truths.

UAP sightings are common and widespread.

In compelling testimony before the panel, retired Cmdr. David Fravor and former Navy fighter pilot

Ryan Graves both boldly recounted their terrible experiences with unidentified aircraft.

With unflinching conviction, Graves recalled a terrible encounter that happened off the coast of Virginia Beach in 2014. In the F-18's cockpit, he saw an enigmatic object that was unlike any machine he had ever seen: a "dark grey or black cube ensconced within a clear sphere," with a diameter of five to fifteen feet. It seems inconceivable how the UAP managed to remain stable in the face of strong gusts equivalent to hurricanes.

Despite the gravity of the situation, Graves' squadron filed a safety report swiftly, only to be greeted with an emphatic silence from official channels. His unflinching statement aligned with the perception of many local pilots that these kinds of encounters with UAPs were not unusual, but frighteningly frequent.

The Official Government Acknowledgement of Meetings with Alien Visitors

Driven by the pursuit of safety and the truth, Graves established Americans for Safe Aerospace, an unwavering bastion of support for those pilots who dared to report UAP sightings. He unwaveringly said that the enigmatic objects seen by military and commercial pilots fly beyond the current technical capabilities of our nation and beyond any rational explanation.

"If the sensor and video data I witnessed were made public, the very foundation of our country's discourse would undergo an unprecedented transformation," he said with a sudden certainty.

"Amazing technology" that is not available to us

Retired Navy Cmdr. David Fravor told the panel about a scary encounter with a UAP that was

The Official Government Acknowledgement of Meetings with Alien Visitors

captured on camera in 2004 and released to the public by the Pentagon in 2020.

Consider this: Fravor and three other military personnel stared in disbelief as a white, "Tic Tac"-shaped flying object appeared above the Californian city of San Diego. No rotors are present. rotor wash of zero. It's an interesting unmanned aerial vehicle (UAP) that breaks all established aviation laws; it has no wings or apparent flight control surfaces. It defied the rules of physics, accelerated at an unimaginable velocity, and disappeared before their eyes, leaving no trace of disturbance, as they attempted to approach this amazing miracle.

Consider this: The technology they met was considerably more advanced than anything that humanity has hitherto encountered. A technology so amazing and unattainable that Fravor had to declare it inconceivable.

The Official Government Acknowledgement of Meetings with Alien Visitors

To be clear, Fravor is not a UFO fanatic. Nevertheless, with four sets of eyes, they saw a strange and confusing reality that defied all assumptions. An unparalleled and perplexing technical prowess.

But throughout the years, authorities did not do anything to address these extraordinary events, leaving Fravor and his fellow pilots in the dark and without an explanation. This narrative is an unambiguous testimony to a confusing reality and a pivotal moment that requires our focus and unwavering pursuit of answers. The exchange made a lasting impression, the evidence is unquestionable, and the silence is unbearable. It is time to confront the enigma that hangs over our sky.

There are amazing gasps coming from the overflow room!

Look, the much awaited hearing was at last open to the public, and throngs of fervent supporters risked hours-long lineups to get seats within its hallowed walls. Among the more than a hundred enthralled spectators, a determined 22-year-old from New York City proclaimed his attendance at this historic occasion. He requested anonymity due to the ongoing stigma associated with the subject. He knew full well that history may collapse before his own eyes.

He, along with the rapt crowd, waited with great expectation as well-known war veterans like Grusch, Graves, and Fravor related their amazing experiences. He proclaimed his unwavering belief in the veracity of these three unwavering witnesses' accounts in the face of doubt, even though he acknowledged that, when taken out of context, their remarks may look ludicrous.

But it wasn't just him; the whole audience gasped when Grusch brought up the subject of non-human

The Official Government Acknowledgement of Meetings with Alien Visitors

biologics. An equally dramatic reaction erupted from the rapt audience as Grusch detailed the personal retaliation he endured.

Why now, exactly?

I know you're thinking, "Why now?" In the interest of national security, Congress is steadfastly requesting—through this hearing—that the intelligence community provide an explanation for Unidentified Aerial Phenomena (UAPs).

Democratic representative from California, Rep. Robert Garcia, asserts emphatically that "UAPs, whatever they may be, may pose a serious threat to our military and our civilian aircraft, and that must be understood." He ardently believes that more reporting on UAPs is better than decreased, since information fosters safety.

The Official Government Acknowledgement of Meetings with Alien Visitors

Grusch, Graves, and Fravor all vehemently reiterate similar points of view, earnestly calling for the removal of the stigma attached to UAP complaints in order to inspire others to come forward and ardently wanting a "safe and transparent" centralised reporting procedure.

Remarkably, the well-known former Navy pilot Graves reveals that the Office of the Director of National Intelligence and the All-Domain Anomaly Resolution Office only get a pathetic 5% of UAP sightings.

He makes a strong but clear appeal, "I implore us to set aside stigma and face the security and safety risk this topic represents." The security of the country is seriously threatened if UAP are foreign drones. In the event that it's not, a scientific issue exists. In any case, unidentified objects are dangerous for flight safety. America's citizens are entitled to know what's happening in the sky. It is long overdue.

The Official Government Acknowledgement of Meetings with Alien Visitors

The Official Government Acknowledgement of Meetings with Alien Visitors

CHAPTER SIX

FUTURE PROSPECTS AND THE IMPACT ON SOCIETY

After the government officially acknowledged the existence of extraterrestrial encounters, the globe found itself on the cusp of an unprecedented paradigm shift. Chapter five of "UFOs: The Government's Official Acknowledgment of Alien Visitors" looks at the implications for society and what lies ahead. This chapter discusses the many implications of acknowledging the presence of extraterrestrials, ranging from considering the implications of interstellar diplomacy to doubting our place in the cosmos.

The Impact of Official Recognition of UFOs on Public Perception

The official recognition of UFOs by governments throughout the globe sent shockwaves. claims of UFO sightings were either scoffed at or written off as hoaxes and false claims for a long time. With the enigma solved, mankind now had to consider the idea that we weren't the only beings in the vastness of space. In the face of growing public curiosity, the scientific community was compelled to reevaluate its stance on alien life.

The media had a vital role in shaping public perception, disseminating data, and initiating conversations around the significance of the freshly unearthed revelation. The combination of cynicism and amazement led to divergent opinions about the intentions and potential capabilities of extraterrestrial encounters. Philosophers,

The Official Government Acknowledgement of Meetings with Alien Visitors

theologians, and scientists reexamined long-held beliefs about what makes humans unique and our place in the universe in intense disputes.

It became the responsibility of governments to provide accurate information while averting misunderstandings and anxiety. In an attempt to inform the public about the nature of the encounters and the continuing research, emphasis was placed on the need for a cool-headed and methodical approach to comprehending this reality that is shifting paradigms.

Government involvement in SETI

The Search for alien Intelligence (SETI) missions were revived with the official government acceptance of alien visitation. Governments who were keen to explore the potential for contact with

The Official Government Acknowledgement of Meetings with Alien Visitors

alien societies increased the amount of money and support they provided to scientists and groups working to find intelligent broadcasts from space.

Collaboration between governments and the scientific community has created new avenues for research and advancement in the area of SETI. The goal of developing sophisticated radio telescopes and other state-of-the-art technology is to investigate the universe for any signals that might indicate the existence of intelligent life beyond Earth.

Even though the hunt for alien intelligence remained a challenging endeavour, the government's involvement provided the field respectability and inspired a new generation of scientists to actively join in this huge cosmic quest.

The Official Government Acknowledgement of Meetings with Alien Visitors

Executive Orders Concerning Extraterrestrial Relations

Interacting with alien cultures raised serious ethical, diplomatic, and security problems. As governments began formulating policies and procedures to handle possible contact situations, they established international committees to assist them in navigating this uncharted territory.

The primary ethical considerations were respect for extraterrestrial civilizations and safeguarding Earth's interests in the case of contact with highly evolved aliens. Ensuring harmonious interactions and avoiding needless damage to both parties become top priorities.

Diplomatic problems emerged as states argued about who should represent humanity in interplanetary talks. Nations considered uniting to

The Official Government Acknowledgement of Meetings with Alien Visitors

highlight to potential extraterrestrial visitors the values and goals that our planet shares.

A rising number of governments were concerned about security, therefore they evaluated their military plans and emergency plans. There have been debates over the benefits and drawbacks of disclosing certain secret information, balancing the right of the public to know versus national security concerns.

The official admission by governments of alien visitations had a profound effect on public opinion, sparked scientific inquiry, and raised difficult issues about humanity's role in the universe. Governments attempting to navigate this unknown territory faced a combination of possibilities and problems in the face of possible contact with alien civilizations. In the next chapters, we will resume our journey as we explore the complexities and unpredictable nature of interstellar interactions and search for answers

The Official Government Acknowledgement of Meetings with Alien Visitors

to the perplexing issues that arise outside of Earth's jurisdiction.

The Official Government Acknowledgement of Meetings with Alien Visitors

www.ingramcontent.com/pod-product-compliance
Lightning Source LLC
Chambersburg PA
CBHW071607270726
48661CB00019B/1631